Gabriela Lagonegro

Good Manufacturing Practices for Fishery Products

Gabriela Lagonegro

Good Manufacturing Practices for Fishery Products

ScienciaScripts

Imprint
Any brand names and product names mentioned in this book are subject to trademark, brand or patent protection and are trademarks or registered trademarks of their respective holders. The use of brand names, product names, common names, trade names, product descriptions etc. even without a particular marking in this work is in no way to be construed to mean that such names may be regarded as unrestricted in respect of trademark and brand protection legislation and could thus be used by anyone.

Cover image: www.ingimage.com

This book is a translation from the original published under ISBN 978-613-9-43995-9.

Publisher:
Sciencia Scripts
is a trademark of
Dodo Books Indian Ocean Ltd. and OmniScriptum S.R.L publishing group

120 High Road, East Finchley, London, N2 9ED, United Kingdom
Str. Armeneasca 28/1, office 1, Chisinau MD-2012, Republic of Moldova, Europe
Printed at: see last page
ISBN: 978-620-6-53179-1

SUMMARY

In this final work proposed in the National Auditor's Degree of Food Industries of the National University of the Centre of the Province of Buenos Aires, a Manual of Good Manufacturing Practices of a Fish and Seafood Processing and Elaboration Cold Storage Establishment is developed.

Based on the data obtained on the establishment and the information collected on Good Manufacturing Practices, an instruction manual was devised to implement this prerequisite.

This establishment is located in the port of Ingeniero White near the city of Bahía Blanca, and is mainly dedicated to the procurement, processing, preservation and sale of fish and seafood.

The current global problem indicates that 80% of Foodborne Transmissible Diseases (FBDs) are linked to problems of poor handling and hygiene practices in food processing sites, as well as the actions of operators who are not properly trained to perform their work in a responsible manner and company owners who only want to produce large quantities without taking into account the safety and quality of their products.

It is essential to comply with the requirements of our country's regulations by implementing "Good Manufacturing Practices", with the primary objective of achieving safe and good quality food.

ACKNOWLEDGEMENTS AND DEDICATIONS

First of all, I would like to thank my family, especially my children Catalina, Candela, and Geronimo, who always encourage and support me in each of my projects.

To Dr. Héctor Pettinato who passed on to me the pleasure of transmitting knowledge to achieve safe and quality food, both in my undergraduate career and in this postgraduate course.

I dedicate this work to the memory of two great men who had a great impact on my personal and working life: my father Ricardo Lagonegro and Don Domingo Caserma.

CHAPTER 1 INTRODUCTION

In order to achieve excellence and to be able to compete in today's markets, every company's primary objective must be to ensure the QUALITY of its products.

The initial step to achieve this is the implementation of GOOD MANUFACTURING PRACTICES (GMP). These are referred to as pre-requisite programmes together with Standard Operating Procedures for Sanitation. They are programmes that ensure appropriate operational and environmental conditions for the production of safe food. These programmes provide the operational and environmental conditions necessary for the production of safe food. They are hygiene and handling procedures that are the basic requirements for market participation, ultimately leading to safe and wholesome food.

The premise of these programmes is to develop, document, implement, monitor and control the programmes that are prerequisites for the implementation of HACCP.

To achieve this, it is essential that all actions are documented, and that responsible persons are involved in each task.

Current regulations include the entry into force of Common Market resolution 80/96, which establishes a regulation for compliance with GMP and hygienic-sanitary conditions in food processing and industrialising establishments. This regulation is based on parts of the US Code of Federal Regulations and general principles of hygiene of the *Codex Alimentarius*, as well as the *Codex* Committee on Food Hygiene, Service Order 6/98 and SENASA Resolution 238/98.

In the case of Argentina, Chapter II of the Argentine Food Code and

Decree 4238/68 mention these programmes.

Presentation of the Establishment

Establishment N° Official 1518 "PESQUERA COSTAS y MARES".

Location: Avda. Las Oreas and Los Cazones - (8106) - Ing. White - Partido (department) of Bahía Blanca - Province of Buenos Aires - Argentina.

Activities performed: reception, filleting/peeling and packaging of fresh and frozen fish and seafood.

Proprietary Trading Firm:

"PESQUERA COSTAS y MARES" by AGUSTÍN MATEO

Location of the industrial plant: the plant is located in the vicinity of the Port of Ing. White, an industrial settlement area, in the area between Las Orcas, Los Delfines, Los Cazones and Los Corales streets.

It has the following cadastral designation:

District II, Section C, Block 952, Plot 2b; 3c; 12b; 27; 28; 1d

This establishment is located in non-flooded areas, where the settlement of this type of plants is permitted. Most of the raw material processed in the establishment is captured in canoes and own boats in the maritime littoral that goes from Riacho Azul, Ría de Bahía Blanca and Villalonga. Raw material from other authorised establishments is also processed.

Corporate Organisation Chart

Organisational map:

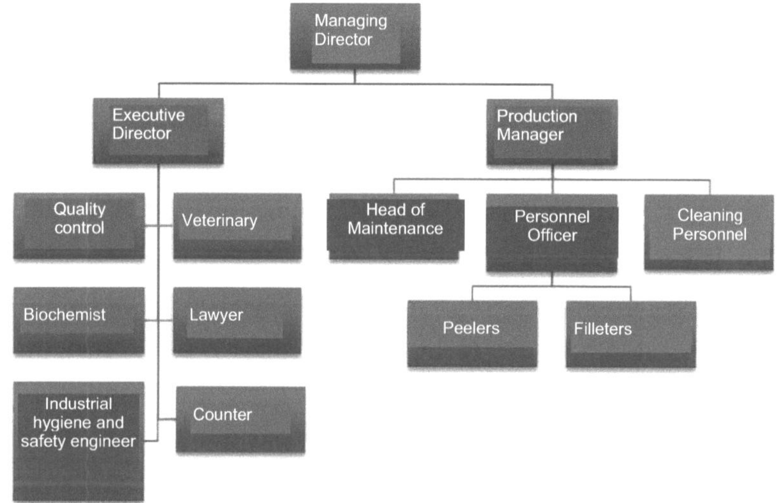

<u>Company description:</u> "Pesquera Costas y Mares" is a family company, whose capital is national. It employs a number of professionals hired as service providers, who report directly to the Executive Director. On the other hand, the skilled labour force depends on the production manager, who is hired directly by the company, while the peelers and filleters are employed by a work cooperative. Its facilities are located in the city of Bahía Blanca, in the port of Ingeniero White, in the province of Buenos Aires. This is where the fish and seafood refrigeration and processing plant is located (Item XLVI).

It processes shrimp and prawns, as well as some species of fish such as gatuzo, whiting, sole, silverside, etc., which are supplied by its own and third-party canoes and boats.

The commercialisation is carried out within the Argentine Republic, being transported by our own transport or by contracting third parties if necessary.

Number of staff: the plant employs 5 women and 15 men on the permanent staff and around 100 people who belong to a Worker Cooperative, whose work is temporary and coincides with the shrimp and prawn catching season.

Working hours: from 5hs am to 17hs pm, this timetable is subject to the entry of canoes and boats whose return depends on the tidesWorking hours: Filleting from 5hs to completion (generally 10hs), shrimp peeling from 5hs to 13hs, with a fifteen minute break every two hours.

The plant operates from Monday to Saturday, on Sundays the staff on duty receives goods from boats and canoes.

The administrative area works from 8 a.m. to 3 p.m. from Monday to Friday.

The company has external advisors and professionals such as accountants, industrial hygiene and safety engineers, veterinary surgeons and lawyers who provide their services to the company.

Type of qualification:

It is an Industrial Establishment, with SENASA authorisation, for item 36 (fish and seafood processing), dedicated to the processing of raw and cooked, whole or peeled, chilled or frozen prawns and shrimps in different presentations according to the demands of the buying markets.

The raw material comes from the company's own vessels or from other nationally licensed companies.

Content

1. OBJECTIVES AND SCOPE OF APPLICATION:

This Manual of Good Manufacturing Practices has been drafted with the aim of establishing hygienic-sanitary standards to be applied at the various stages of product processing.
It will be constantly applied in all areas of the company in order to ensure the highest product quality and thus provide assurance to current and potential customers of the company.

- To obtain foodstuffs of excellent quality and adequate hygienic handling, which do not constitute a risk to the health of the consumer.
- Achieve predictability of Foodborne Diseases (FBDs) and reduce their incidence. To reduce the risks of disease transmission due to operational failures.
- To train food handlers both in theory and in practice in order to obtain safe and high quality products, so that each member of this industry becomes aware of their responsibility when handling food.
- Ensure that the processing of fish and shellfish is only carried out in approved plants intended for this purpose, discouraging the peeling of shrimp in private homes or in non-approved places.

- Incorporate hundreds of people into the labour market by providing them with decent work in suitable environments.

CHAPTER 2 DESIGN AND CONSTRUCTION OF THE PROCESSING FACILITIES

Buildings and facilities

Plant and its surroundings

Areas adjacent to the plant that are under the control of the Company are maintained in a condition that protects against food contamination. The methods used to maintain adjacent areas are:

- Storage of equipment, proper removal of debris, and trimming of grass and weeds that may be an attraction, breeding place or harbouring place for pests.
- Maintenance of streets, yards and parking places so that they do not constitute a source of pollution.
- Maintenance in places suitable for drainage which in a way may contribute to contamination of foodstuffs through infiltration, or mud carried by shoes, or which provides a niche for pests.
- Waste treatment and disposal systems shall be operated in an appropriate manner so that they do not constitute a source of contamination.
- Land surrounding the plant that is outside the control of the Company and is not maintained in the manner described above; for these cases, greater care is exercised within the plant by means of inspections, pest control and any other means to exclude dirt that could be a source of contamination for foodstuffs.

Design and construction features of the plant

The plant building and structure are of a size, construction and design that facilitates its maintenance and sanitary operations to serve its purpose in food processing.

Both the equipment and the plant layout used in the design and construction favour the execution of the processing activities as well as the hygiene and disinfection of equipment and its correct maintenance.

This plant is built entirely on a reinforced concrete structure, it is constructed of masonry, with tiled areas and others covered with tongue-and-groove PVC sheets. It has a covered area of 1413 sqm.[2]

The plant and its structures have sufficient space for the installation of equipment and storage of materials necessary for the maintenance of sanitary operations and the production of a safe food product.

Precautions are taken to reduce contamination of the food, food contact surface, or packaging material against microorganisms, chemicals, debris, and foreign matter. The potential for contamination is reduced by safety controls and appropriate operating practices, or effective design, including separation of the operation by which contamination is likely to occur, by one or more of the following methods: location, timing, partitioning, airflow, or other effective means.

Transit and storage areas for fishery products are slip-resistant, waterproof, easy to clean and disinfect, with easy evacuation of water, liquid drains are siphoned, open chutes carrying liquids have a stainless steel sieve cover, which allows for easy inspection. Floors, walls and ceilings are constructed in such a way that they can be adequately cleaned, kept clean and in good condition: leaks or condensation on pipes and walls are unobstructed and there is sufficient space to allow employees to carry out their tasks and protection without contamination of food, food contact surfaces, clothing or through personal contact.

The majority of the floor is tiled in white enamelled San Lorenzo type tiles with a grout joint, in the chambers, white sanitary panels with pre-painted, the meeting angles are rounded and easy to clean.

Lighting is adequate in hand washing, boot washing, changing rooms and toilets and in all areas where food is inspected, processed, or stored and where equipment and utensils are washed.

The ceilings are made of white plastic panels, which are easy to clean, and go through pipes, electrical conduits, refrigeration pipes, etc.

Ventilation is adequate in areas where odours and/or vapours may contaminate food. There is a system of air supply in a manner that reduces the potential for contamination of food, packaging material and food contact surfaces.

Window sills with a sufficient slope to facilitate cleaning and ensure that dust and dirt do not accumulate.

All openings to the outside are fitted with mosquito netting to prevent flying insects from entering.

The connecting doors are made of anodised aluminium with watertight polycarbonate plates. In the chambers they are made of stainless steel. In bathrooms and changing rooms, doors are made of white plates.

Health operations

General maintenance

The building and other physical plant facilities are maintained in good repair and sanitary condition to prevent food contamination. Utensils and equipment are washed and sanitised in a manner that protects against contamination of food, food contact surfaces and packaging material.

Substances used for cleaning and disinfection

Storage of toxic products.

Detergents and disinfectants used in cleaning and sanitation procedures are free of micro-organisms, safe and efficient for their intended use. They meet this requirement and can be verified by any effective method including purchase of these substances under vendor certification. The following materials are only used and stored in the plant where the food is manufactured or where the food will be exposed:

- . Those required to maintain clean and sanitary conditions.
- . Those required to be used in the laboratory for test analysis.
- . Those that are necessary to be used in the different stages of the process.
- . Chemical detergents, disinfectants and insecticides are identified, maintained and stored in a manner that prevents contamination of food, food contact surfaces and food packaging material.

Cleaning of food contact surfaces

The contact surface used for processing, or for holding food is dry and in a sanitary condition for the time it is to be used. Those surfaces that need to be wet cleaned, where necessary, shall be disinfected and dried before use.

*. When cleaning during processing, it is necessary to protect the food against the introduction of micro-organisms, all food contact surfaces should be washed and disinfected before and after each interruption of work during which they may have been contaminated.

*. Surfaces that do not come into contact with food, equipment used

in the plant are cleaned as often as necessary to avoid contamination of the food.

Pest and contaminant control

Pests

All areas of the plant are enclosed with polycarbonate windows protected by screens or fixed windows. Doors and windows are hermetically sealed. A pest control plan (disinsectisation, disinfection and rat extermination) is in place. The plant has an Integrated Pest Management Manual.

Dust

This is prevented by paving the loading and unloading areas and the access road to the raw material intake area. In addition, the exit of the finished product is located on a paved road.

Building features to avoid cross-contamination

The verification of the process flow diagram on the plant plan, together with the Sanitation Standard Operating Procedures (SSOPs) for the product processing room, ensures safety from the point of view of cross-contamination. Thus the facilities allow operations to be carried out under hygienic conditions from the arrival of the raw material to the production and storage of the finished product.

Aspects linked to the food handling area

The processing room has a smoothed, coarse-grained, non-slip cement floor, the walls are made of reinforced concrete and the ceiling is covered with insulating material. In most of the plant, white enamelled San Lorenzo type tiles are used with a grout joint, in the chambers, white sanitary panels with pre-painted, the corners are rounded and easy to clean.

The floor has reinforced concrete covered gutters lined with glazed ceramic on the inside, these open gutters that transport liquids have a stainless steel sieved cover, which allows easy inspection.

The joints between walls and floors form a rounded sanitary plinth that facilitates cleaning and disinfection.

The transit and storage areas for fishery products are non-slip, waterproof, easy to clean and disinfect, with easy evacuation of water, and the liquid drains are siphoned,

Toilets, boot-washing facilities, changing rooms and staff toilets

Both the toilets and the changing rooms are separate from the production and processing areas, and are located completely apart from the rest of the plant, with no direct connection between them and the other facilities. The access of personnel in street clothes to the changing room area is independent from the access to the processing area. Personnel entering the processing area from the changing rooms must pass through sanitary filters.

Location of inputs, raw materials and finished products

The raw material enters through a separate entry for this purpose.

The sector is enclosed by hermetically sealed stainless steel doors and is physically separated from the processing room by a curtain of 60 % overlapping PVC strips.

The other primary and secondary inputs enter through the finished product gate at times that do not coincide with the output of finished product, in order to avoid cross-contamination, and are deposited in the input sector.

Water supply

Description of drinking water supply

The plant is supplied with water from the public water supply.

Therefore, for both processing and manufacturing, water provided by the drinking water supply company of the city of Bahía Blanca is used. It is also supplied by a 420 m deep semi-surge well, cased, with a small reception tank and centrifugal pump for distribution to places of use.

Distribution is through stainless steel pipes.

The storage is done in a cistern suitably constructed for this purpose. The water is pumped to a material tank from where it is distributed to the entire plant through PVC and galvanised iron pipes.

Tank/cistern cleaning is carried out every six months, or when deemed necessary, in the event of contamination of the tank/cistern from any other cause. The cleaning procedure is detailed in the Company's Sanitation Standard Operating Procedures Manual (SSOP).

To ensure the potability of the mains water, an automatic pump-type chlorine dosing system has been installed between the cistern and the water tank, which ensures a constant dosage. This dosage is unique for the whole plant.

A chlorination protocol is in place, where the automatic chlorinator, located upstream of the tank with volumetric and flowmeter alarm systems, with sound and light warnings. The water will be maintained with a free chlorine level of 0.2 to 0.5 ppm at peak usage. Chlorine is monitored on a daily basis and the results are recorded in a spreadsheet. Physical-chemical and microbiological analyses of the water are carried out on a monthly basis and are sent to the Veterinary and Bromatological Analysis Laboratory of the SENASA network. These results are filed in the Quality Control office in the folder labelled "Water".

Drinking water system connections do not intersect with any non-potable water connections.

Means of information used

The Competent Bodies reporting chemical and physico-chemical specifications are as follows:

Argentine Food Code with Mercosur regulation SENASA, Servicio Nacional Sanidad Agroalimentaria.

Use of non-potable water

All the water used by the company is potable, so there is no problem of confusion between non-potable and potable water.

Gaseous effluent and wastewater disposal

Water vapour is generated as a gaseous effluent from the firing room, which is removed by air forcers located in the upper part of the room. Vapours are also generated in toilets and changing rooms which will also be removed by mechanical means.

In both cases, the aim is to ensure a vapour-free environment to avoid condensation on walls and ceilings.

The liquid effluents generated in the plant come from the washing of countertops, utensils, floors and drawers.

Effluent from toilets and changing rooms, also known as sewage, is collected by a network of PVC pipes and discharged into a collection pipe outside the building and discharged into the sewage system.

Once collected, drains from basins, gutters and floor gratings will pass through an inspection chamber which is also used as a sampling site for submission to the laboratory.

The plant is criss-crossed by drainage gutters made of reinforced concrete, lined with glazed ceramic on the inside, these open gutters that transport liquids have a stainless steel sieved cover, which allows for easy inspection.

They flow into a collector channel made of the same materials, which carries the waste water to a system of settling chambers outside the plant. The settling chambers are made of reinforced concrete. When necessary and at least once every 60 days, the solid waste is removed and the settling chambers are cleaned (Empresas "Los Campanelli" SRL).

Sewage from toilets and changing rooms is collected by a network of PVC pipes which discharge into a central black painted cast iron collection pipe. Dining room and office toilets are collected in the same way.

Changing rooms and toilets

The changing rooms and toilets are separate from the processing area. They are separated and identified for both sexes. To facilitate cleaning, they have a non-slip ceramic floor and the walls are covered with white ceramic tiles up to the ceiling. The lighting fixtures have a high-density polyethylene base and polycarbonate cover. Fluorescent tubes are used. For ventilation there are partitions made of aluminium and glass. The sanitary facilities have toilets, urinals and washbasins. The bathrooms have liquid soap for hand washing and disposable paper for drying hands. The receptacles for the disposal of used disposable towels consist of plastic containers with swinging lids.

Cleaning and disinfection facilities

Specially designed facilities are available for this purpose, but in order to ensure that cleaning utensils are easily and quickly accessible, they are placed on hangers in each sector of the plant.

Detergents and disinfectants used in cleaning and sanitation procedures are safe and effective for their intended use. The purchase of these substances is under vendor certification. All products used in the plant are identified, maintained and stored in a correct way, having a sector for all cleaning products and another one for toxic products. This prevents contamination of food, food contact surfaces and food packaging material.

Storage of waste and inedible material

The waste resulting from the activities carried out in the different sectors is placed in colour-coded high-density polyethylene bins. They are removed from the corresponding sectors to be transported by means of a stainless steel trolley, to the trunk that connects to the waste container, which is located inside the plant, completely separate from the other sectors.

To control the entry of pests, baits are placed around the perimeter of the building.

The plant's waste consists of:

1 Disposable towels at hand-washing facilities, which are placed in the receptacles provided. These are emptied at the end of the working day.

2 Paper and polyethylene packaging from primary packaging, which is discarded in crates to be taken to the waste disposal sector and placed in the container.

3 Waste from the canteen, offices, toilets and changing rooms, which is handled by the cleaning staff assigned exclusively to these sectors, collected in high-density plastic bags and deposited at the end of the day in the corresponding container.

Return of products

In the event of a refund, the following considerations shall be taken into account:

a) the product is fit for consumption but does not meet the customer's specifications.

b) The product has undergone microbiological or physico-chemical deterioration rendering it unfit for consumption.

Taking these items into account, proceed as follows:

For case a) the returned product will be destined for markets with different requirements.

In case (b), confiscation shall take place.

The competent official body (SENASA) shall be notified in any case.

Equipment and utensils

Materials

The equipment and utensils used in the plant are made of stainless steel and innocuous materials, which do not transmit toxic substances, odours or tastes, are non-absorbent, and resist both corrosion and repeated cleaning and disinfection operations. The equipment has smooth surfaces that facilitate cleaning, although it is impossible to avoid the existence of grooves and crevices that are inherent to the design. It is for this reason that the steps detailed in the SSOP Manual must be carefully followed to ensure effective cleaning and to control sources of contamination.

The equipment is installed and maintained in such a way that it is easy to clean and has space around it to allow for proper manoeuvring.

The food contact surface is corrosion resistant when in contact with food. This surface is constructed of a non-toxic material and designed to withstand the environment in which it is used and the reaction of the food, and when detergents and sanitising agents are applied for cleaning. The food contact surface is maintained in a manner that protects the food from being contaminated by any source.

Design and construction

Information on equipment and utensils

The equipment used is of a design and construction that facilitates and ensures thorough cleaning and disinfection. In general, all parts are visible so that efficient inspections can be carried out.

The bonds on the food contact surface have a smooth or maintained bond in a way that reduces the accumulation of food particles, dirt, organic particles and reduces the opportunity for the growth of undesirable micro-organisms.

Chillers and freezers are properly identified with name and number. They are used to store and keep foodstuffs that are capable of encouraging the growth of micro-organisms, have a thermometer and a thermograph for recording the temperature, installed in such a way as to show the exact temperature inside the chamber, and in turn this apparatus is fitted with an automatically controlled temperature regulator or with an alarm system which indicates temperature changes. (Data TESTO).

Ice house capacity / Ice factory - Ice silo / Temperature recording

Fresh raw material reception chamber

Capacity: 15 tonnes

Frozen product storage chamber: 3 (three)

Capacity: 250 tonnes for shrimp and prawns

Tunnel freezer:

Approximate capacity per round: 8 tonnes

Freezing time: 9 hours

Ice factory: 2 (two)

Capacity 10 tonnes per day each

Ice silo:

Capacity: 20 tonnes

Containers for inedible materials and wastes

The containers used for the plant's solid waste are made of plastic material (non-absorbent), easy to clean and easy to dispose of the contents. Their structure ensures that they do not leak. They are appropriately marked in black and are only used for this purpose. They are engraved with the name of the company.

Measuring and recording instruments

The instruments used for measuring and recording are: digital meat thermometers, alcohol thermometers for recording at room temperature. The plant has a standard thermometer which is calibrated once a year at the National University of the South. This thermometer is used to calibrate the different thermometers used in the plant twice a year.

Temperature recording:

Mini data logger brand TESTO with integrated display and memory for 3900 readings, origin Germany. Programmed to measure in 2 (two) hour intervals. Reading of records every 15 (fifteen) days, with the possibility of reading at the time requested.

There are 3 (three) instruments located in each of the frozen product storage chambers.

In support of this computerised record, the waiter on duty makes a tour every hour, making a manual entry on a Temperature Log sheet. Digital and clock scales are also used, which are calibrated once a year against the standard weight, and checked by spreadsheet on a monthly basis.

CHAPTER 3 PERSONAL HYGIENE AND HEALTH REQUIREMENTS

Hygiene education

The plant's quality control department permanently instructs all staff on food hygiene and safety standards. Training is given directly while working and periodic refresher courses are also given by the Head of Quality Control. The content of the training and refresher courses given to personnel is recorded in their respective folders, attaching a sheet signed by those attending the courses and an eventual multiple choice evaluation in written form.

3.1. Health status

The Argentinean Food Code establishes in its regulations that all individuals involved in the processing of fresh and/or frozen fishery products, from those in charge of loading and unloading the raw material and the finished product, to those in charge of handling, and even the staff of the food processing plant, must prove that they are fit to work with foodstuffs.

The health care of the permanent staff is carried out by the "Empresa de Medicina Laboral Protección Privada EML", which is responsible for certifying the fitness of the staff and annually submits the workers and executives of the plant to a complete check-up consisting of clinical examination, blood tests, X-ray, audiometry, and ECC.

The workers of the worker cooperative annually renew their Occupational Health Document at the Hospital Municipal de Agudos Dr. Leónidas Lucero in the city of Bahía Blanca.

Communicable diseases and injuries

Workers suffering from any illness (cold, conjunctivitis, diarrhoea, cough, etc.) or with any injury to their hands or arms must inform the Personnel Manager. This worker will be removed from the production line to consult a doctor if necessary, and may return to work once he/she has recovered from the illness. Any wounds should be covered with a dressing and gloves should be worn for protection.

In the event that the injury was caused during the working day, the procedure designed by the ART will be carried out.

Hand washing

Employees will be trained on how and when to properly wash and disinfect their hands.

All persons involved in the processing should always wash and disinfect their hands, at the start of the task, after using the toilet, after handling contaminated material and whenever they otherwise leave the task.

Hands shall be washed with warm water and a detergent-disinfectant solution.

Employees will be trained on how and when to properly wash and disinfect their hands.

Personal hygiene

The hygiene of the personnel working in the plant is of utmost importance as human beings are a reservoir of micro-organisms and can be a vehicle for diseases.

Hand washing is verified before starting work, after a break, after going to the toilet or after any non-processing task.

Before entering the processing area, operators pass through the sanitary filter.

The faucets are knee-operated, using liquid soap from a dispenser and paper towels which are disposed of in a container intended solely for this purpose.

Special care must be taken with maintenance and foremen who

handle cleaning agents, hydrocarbons, corrosives, etc. During production hours, they may only access the processing sectors in case of emergency and with the authorisation and in the presence of the supervisor.

Personal conduct

Responsibilities of the Processor:

It is the responsibility of the Processor to designate within the Plant the personnel in charge of establishing the procedures that ensure that the products produced, deposited and/or fractionated in the plant are harmless, and therefore do not constitute a risk to the health of the consumer, for which the following guidelines must be taken into account:

- Establish production standards following Good Manufacturing Practices and Standardised Sanitation Operating Processes.
- Develop these instructions (manuals) in writing
- Provide measuring and/or recording instruments duly calibrated with national and international standards.
- Document in spreadsheets the actions taken
- Monitor and control reception, processing, warehousing and dispatch processes.
- Verify and review procedures and documentation
- Records documenting compliance with the described procedures shall be kept and made available to the health authority (veterinary inspection, external auditors and buyers).
- Implement training courses for all plant operators.

Operator responsibilities

- Perform duties in compliance with production standards

following Good Manufacturing Practices and Standardised Sanitation Operating Processes.

- Become aware of the rules
- Use the instruments to measure and/or record properly
- Document in spreadsheets the actions taken
- Monitor and control reception, processing, warehousing and dispatch processes.
- Attend training courses
- Report any abnormalities to your manager.
- To take responsibility for safe products, which do not constitute a health risk for the consumer.

Prohibitions:

- Smoking inside the plant, eating and/or drinking in the processing room, chewing gum or eating sweets during processing is prohibited.
- No salivating or other acts which may impair the hygiene of the food or the establishment.
- Make-up, nail polish, perfume, personal effects such as rings, watches, bracelets, etc. are forbidden.
- These prohibitions are applicable to all personnel working in the plant, and there is the possibility of being sanctioned in the event of any infringement of these prohibitions.

REGISTRATION AND MONITORING

The **personnel control sheet** is carried out twice a day. The first recording is made in the morning before the break and the second after the break.

The Personnel Officer shall be responsible for the preparation of this spreadsheet.

ANNEX

FORM N°1 PERSONNEL CONTROL REGISTER

DATE: ....

	PRE OPERATIONAL	OPERATIONAL	REMARKS
	Time:	Time:	
Clothing			
Full uniform			
Cleanliness and condition			
Washing of aprons			
Personal hygiene			
Hand washing			
Presence of personal elements			
Use of sanitary filter			
Health status			
Presence of wounds			
Presenceof Secretions			
Respiratory illnesses (cold, cough, sneezing)			
Digestive diseases (daily) vomiting)			

CORRECTIVE ACTION:

A: ACCEPTABLE F: FAILURE

Signature and clarification of the person in charge

Gloves

If gloves are used, they must be kept clean and hygienic and in undamaged condition. They shall be made of material that allows easy sanitisation, which shall be carried out with the same frequency and under the same conditions as hand washing.

Visitors

All persons visiting the plant must comply with the same rules as those established for the personnel who work there. They must therefore comply with the provisions of this manual and must also wear overalls, caps or hats and boots.

Supervision.

Supervision of all the above points is the responsibility of the Plant Supervisor; in his absence, the Quality Control Manager will be in charge.

Hygiene requirements for processing

Raw material requirements

Establishment's provisions for acceptance of raw material. Personnel in charge of receiving the goods shall not accept any raw material or ingredient containing parasites, micro-organisms or toxic, decomposed or foreign substances that cannot be reduced to acceptable levels by normal sorting and/or preparation or processing procedures.

Raw materials or ingredients must be inspected and sorted before being taken to the processing line and, if necessary, laboratory tests must be carried out. Only clean and sound raw materials or ingredients should be used for further processing.

Raw materials and ingredients stored on the premises of the establishment must be kept in conditions that prevent spoilage, protect against contamination and minimise damage. Adequate stock rotation of raw materials and ingredients must be ensured.

Control of suppliers

Raw materials are purchased from suppliers who comply with agreed specifications, with the aim of ensuring quality and safety of the raw materials.

To this end, a protocol of specifications is drawn up which the supplier must accept and comply with. This will be carried out by the quality control team, which will be the same team that will control compliance. The way to control is by making visits to the supplier companies 2 or 3 times a year, carrying out an exhaustive control of the raw material when it is received at the factory, keeping the registers and the spreadsheets up to date.

Aspects to be specified in the protocol:

- Description of facilities.
- Description of the raw material.
- Intrinsic factors (pH, AW, salt, alcohol)
- Microbiological criteria. Certificates of analysis
- Sampling plans.
- Labelled.
- Storage conditions.
- Description of packaging type, size, quantity.

- Description of the production process, flow chart.
- Requesting documents

  Prevention of Cross Contamination: Effective measures shall be taken to prevent contamination of food material by direct or indirect contact with contaminated material upstream in the process.

Persons handling raw materials or semi-finished products with a risk of contaminating the final product shall not come into contact with any final product until they have removed all protective clothing worn during the handling of raw materials or semi-finished products with which they have come into contact or which have been soiled by raw materials or semi-finished products and have changed into clean protective clothing.

If contamination is likely, hands should be washed thoroughly between handling products at various stages of processing.

All equipment which has come into contact with raw materials or contaminated material must be thoroughly cleaned and disinfected before being used for contact with non-contaminated products.

Potable water is used in the plant. With the approval of the Competent Body. The recirculated water can be used again within the establishment properly treated to avoid conditions such that its use could present a health risk. In these cases the treatment process is kept under constant surveillance.

Elaboration

Level of expertise required by the establishment for plant personnel and operator training: Processing is carried out by trained personnel and supervised by technically competent personnel.

All operations in the production process, including packaging, must be carried out without unnecessary delay and under conditions that exclude any possibility of contamination, spoilage or growth of pathogenic and spoilage-causing micro-organisms.

Containers shall be treated with due care to avoid any possibility of contamination of the processed product.

The methods of preservation and the necessary controls shall be such as to protect against contamination or the occurrence of a risk to public health and against deterioration within the limits of good commercial practice.

STORAGE AND TRANSPORT

Precautions for storage and transport of the finished product

The products produced will be stored in the existing chambers in the plant. There is a computerised stock control system in which the incoming and outgoing goods are recorded.

Precautions to be taken with regard to vehicles used for the transport of finished product

The stored products can be shipped to the customer with our own transport and in vehicles not belonging to the company. In both cases, the vehicles have the corresponding health authorisation for the transport of foodstuffs. An ocular inspection of the hygiene status of the vehicle's box is also carried out and the proper functioning of the refrigeration equipment is checked. The goods leave the establishment with a transit permit issued by SENASA.

Processes of elaboration (operational reports)

Location of inputs and raw materials.

Raw materials enter through a gate, additives enter through the same gate at non-coincidental times, in order to avoid cross-contamination, and are deposited in the input sector. Primary and secondary containers enter directly into the ad hoc warehouse. These sectors are protected by bellows and/or protective overhangs, and are physically separated from the processing room by a curtain of PVC strips.

All supplies are kept in their original packaging and labelling.

Entrance to the processing room

Circulation in the plant

Operators enter the plant from the outside through the La Orcas street access, go up the stairs to the upper floor and enter the separate changing rooms for men and women.

In the front part of the changing room they leave their street clothes and footwear, which are changed for the linen and boots. Once they have finished dressing, they go down the stairs to the shaving area, passing through the sanitary filter where they wash their boots with a brush and sole-washer and then wash their hands.

They are placed around the counters in groups of 8 to 10 operators.

One of them takes a 10kg crate of cooked shrimp and dumps it in the middle of the counter.

The shrimp are peeled by hand. Every 2 hours there is a break of 15 minutes, during which time they go up to the dining room. When they return, they wash their boots and hands and put their aprons back on. At the end of the day they wash their aprons and put them on the coat rack, go up the stairs to the changing rooms, leave their linen and put on their street clothes to leave.

They never return to the processing room in their street clothes.

VISITORS

Visitors adopt the same rules and conditions set out for operators in terms of conduct and clothing.

WORK CLOTHES

The company provides employees with 2 complete sets of equipment per year, while the workers who belong to the peelers' cooperative bring their own clothes. In both cases the clothes are taken to the homes for washing.

The colour chosen for the clothing is white as it allows a quick and easy assessment of the state of cleanliness, including for footwear.

The coverall should be white, if possible without external pockets, made of wash-resistant fabric, as it should be disinfected regularly. White cap or bonnet covering all the hair, including the fringe, as some hair may come loose and fall on the food. The person in charge wears a red cap to be distinguished from the rest of the operators.

Ankle boots and/or rubber boots with non-slip soles

Plastic aprons that are washed before and after use. At break times they are hung on the coat racks.

Maintenance personnel wear blue Grafa® clothing and steel-toed boots. These personnel do not enter the plant during processing.

Laundering of work clothes

At present, the work clothes are washed by the workers themselves, who take the change of clothes used on Fridays and bring them back on Mondays. The company delivers a kit of summer or winter clothes every 6 months.

A laundry room is planned for the coming years to wash the garments in the same plant.

3 STAFF TRAINING

It is the responsibility of the company (manager) to make each and every one of its workers aware of the safety and hygiene standards and to motivate them for the optimum performance of their duties. Achieving product safety and an excellent level of quality.

Training objectives:

- to make operators aware of the importance of their work for the production of safe food and the protection of the health of consumers
- train operators in aspects related to proper food handling
- verify the correct application of the acquired knowledge This

training consists of periodic lectures given by the Veterinarian and by the Safety and Industrial Hygiene Engineer.

Periodically, lecturers are invited (SENASA personnel, personnel from the Pest Control Company, personnel from the Bromatological Diagnostic Laboratory, etc.) or they are encouraged to take courses of interest outside the plant (food handling, food-borne diseases, etc.).

In case of special needs concerning the hygienic handling of foodstuffs, an informal discussion is held in order to communicate the steps to be taken.

In the Training Programme, staff are instructed on:

- Hygiene in food handling. Food contamination. Notions of food-borne diseases.
- Cross-contamination - how to avoid it?
- Correct use of sanitary filters
- Use and maintenance of working clothes

These topics are taught in a clear and practical way. These meetings are held monthly, last 30 minutes, with practical demonstrations on site, followed by multiple-choice questions.

It is taken into account that the plant employs workers with special conditions (hearing impaired, illiterate) who are reinforced with drawings and role imitation. So that all operators have the opportunity to learn, making the participation of each and every one of the operators inclusive.

At the end of the course, each participant receives a certificate of attendance.

Clearly visible reminder signs are placed at key locations.

CONCLUSIONS

The objective of implementing the new working methods incorporating Good Manufacturing Practices, not only leads to comply with the requirements demanded by the competent authorities, but also to reverse social issues by improving the welfare of the population of Ing.

With the construction of plants for the processing and elaboration of fish and seafood, this group of people who in the past, when there were no authorised plants, processed fish and seafood clandestinely in their homes, managed to have a decent job and become part of the food processing chain in our region.

Through the training of handlers, both in theory and in practice, each of these members of the production chain was made aware of their responsibility when working with food.

The education of the workers succeeded in instilling hygiene and health habits, not only at work but also at home. It raised their self-esteem, making them feel useful, valuing their role, and they gained knowledge of different diseases and risks of STDs.

On the other hand, feeling at ease in the place increased loyalty and their work efficiency, and they were able to work in pleasant and harmonious places.

REFERENCES

- Argentine Food Code (CAA)http://www.anmat.gov.ar/alimentos/codigoa/Chapter_VI_2018.pdf Article 270 to 277
- Mercosur (1994) Resolutions of the Southern Common Market Group. Mercosur/GMC/RES. Annex: Technical Regulation on Identity and Quality of Fresh Fish.
- Huss, H.H. (1998) "Fresh fish: its quality and quality changes". FAO Fisheries Technical Paper.
- Updated Regulations for the Inspection of Animal Products. Decree 4238/68. Regulations of the Federal Meat Law. Dr. Héctor Galloni
- Silvestre, A. A. and Rey, A.M. (2004) "*Comer sin riesgos 1*", Buenos Aires, Argentina: Editorial Hemisferio.
- Silvestre, A.A. and Rey, A.M. (2005) "*Comer sin riesgos 2*", Buenos Aires, Argentina: Editorial Hemisferio.
- Food hygiene, microbiology and HACCP. Second edition. Forsythe, S.J. and Hayes, P.R. Editorial Acribia, S.A. Zaragoza, Spain.

APPENDIX - SHEET N°2: TRAINING RECORD OF THE STAFF

Subject of the training: ....

Lecturer:..

Date: ....

SURNAME	NAME	RATING	SIGNATURE

RESPONSIBLE FOR TRAINING:

SIGNATURE AND CLARIFICATION

Table of contents

Printed by Books on Demand GmbH, Norderstedt / Germany